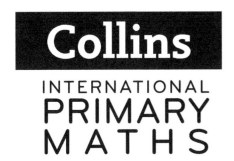

INTERNATIONAL
PRIMARY
MATHS

Progress Book 4

William Collins' dream of knowledge for all began with the publication of his first book in 1819. A self-educated mill worker, he not only enriched millions of lives, but also founded a flourishing publishing house. Today, staying true to this spirit, Collins books are packed with inspiration, innovation and practical expertise. They place you at the centre of a world of possibility and give you exactly what you need to explore it.

Collins. Freedom to teach.

Published by Collins
An imprint of HarperCollins*Publishers*
The News Building
1 London Bridge Street
London
SE1 9GF

HarperCollins*Publishers*
1st Floor Watermarque Building
Ringsend Road
Dublin 4
Ireland

Browse the complete Collins catalogue at
www.collins.co.uk

© HarperCollins*Publishers* Limited 2021

10 9 8 7

ISBN 978-0-00-836960-6

British Library Cataloguing-in-Publication Data
A catalogue record for this publication is available from the British Library.

Author: Peter Clarke
Series editor: Peter Clarke
Publisher: Elaine Higgleton
Product developer: Holly Woolnough
Copyeditor: Tanya Solomons
Proofreader: Catherine Dakin
Answer checker: Steven Matchett
Cover designer: Gordon MacGilp
Cover illustrator: Ann Paganuzzi
Illustrator: Ann Paganuzzi
Typesetter: Ken Vail Graphic Design Ltd
Production controller: Lyndsey Rogers
Printed and bound in the UK using 100% Renewable
Electricity at CPI Group (UK) Ltd

Every effort has been made to trace copyright holders. Any omission will be rectified at the first opportunity.
p56t Boyo Ducks/Shutterstock.

MIX
Paper from
responsible sources
FSC www.fsc.org **FSC™ C007454**

This book is produced from independently certified FSC™ paper
to ensure responsible forest management.

For more information visit: **www.harpercollins.co.uk/green**

The publishers gratefully acknowledge the permission granted to reproduce the copyright material in this book. Every effort has been made to trace copyright holders and to obtain their permission for the use of copyright material. The publishers will gladly receive any information enabling them to rectify any error or omission at the first opportunity.

Cambridge International copyright material in this publication is reproduced under licence and remains the intellectual property of Cambridge Assessment International Education

This text has not been through the Cambridge International endorsement process.

Contents

Introduction

The Progress Books include photocopiable end-of-unit progress tests which are designed to assist teachers with medium-term 'formative' assessment.

Each test is designed to be used within the classroom at the end of a Collins International Primary Maths unit to help measure the progress of learners and identify strengths and weaknesses.

Analysis of the results of the tests helps teachers provide feedback to individual learners on their specific strengths and areas that require improvement, as well as analyse the strengths and weaknesses of the class as a whole.

Self-assessment is also an important feature of the Progress Books as feedback should not only come from the teacher. The Progress Books provide opportunities at the end of each test for learners to self-assess their understanding of the unit, as well as space for teacher feedback.

Structure of the Progress Books

There is one progress test for each of the 27 units in Stage 4.

Each test consists of two pages of questions aimed at assessing the learning objectives from the Cambridge Primary Mathematics Curriculum Framework (0096) for the relevant unit. Where appropriate, this also includes questions to assess learners' development in one or more of the Thinking and Working Mathematically characteristics, as indicated by the TWM star on the page. All of the questions are typeset on triangles to indicate that they are suitable for the majority of learners and assess the unit's learning objectives.

Pages 5 to 9 include a list of photocopiable *I can* statements for each unit which are aimed at providing an opportunity for learners to undertake some form of self-assessment. The intention is that once learners have answered the two pages of questions, they turn to the *I can* statements for the relevant unit and think about each statement and how easy or hard they find the topic. For each statement they colour in the face that is closest to how they feel:

☺ I can do this   ☺ I'm getting there   ☹ I need some help.

A photocopiable variation of the Thinking and Working Mathematically Star is also included in the Progress Books. This version of the star includes *I can* statements for the eight TWM characteristics. Its purpose is to provide an opportunity for learners, twice a term/semester, to think about each of the statements and record how confident they feel about Thinking and Working Mathematically.

Administering each end-of-unit progress test

Recommended timing: 20 to 30 minutes, although this can be altered to suit the needs of individual learners and classes.

Before starting each end-of-unit progress test, ensure that each learner has the resources needed to complete the test. If needed, resources are listed in the 'You will need' box at the start of each test.

On completion of each end-of-unit progress test, use the answers and mark scheme available as a digital download to mark the tests.

Use the box at the bottom of the second page of each end-of-unit progress test to either:

- write the number of marks achieved by the learner out of the total marks possible.

- sign or initial your name to indicate you have marked the test.

- draw a simple picture or diagram such as one of the three faces (☺, ☺, ☹) to indicate your judgement on the learner's level of understanding of the unit's learning objectives.

- write a brief comment such as 'Well done!', 'You've got it', 'Getting there' or 'See me'.

Provide feedback to individual learners as necessary on their strengths and the areas that require improvement. Use the 'Class record-keeping document' located at the back of the Teacher's Guide and as a digital download to update your judgement of each learner's level of mastery in the relevant sub-strand.

I can statements

At the end of each unit, think about each of the *I can* statements and how easy or hard you find the topic. For each statement, colour in the face that is closest to how you feel.

Unit 1 – Counting and sequences (A)	Date:			
• *I can* count on and back in steps of 10, 100 and 1000.		☺	😐	☹
• I can recognise and make generalisations about adding and subtracting combinations of odd and even numbers.		☺	😐	☹
• I can recognise and extend number sequences.		☺	😐	☹
Unit 2 – Counting and sequences (B)	**Date:**			
• I can count on and back in 1-digit steps of constant size.		☺	😐	☹
• I can count beyond zero to include negative numbers.		☺	😐	☹
• I can recognise and extend number sequences.		☺	😐	☹
• I can recognise and extend the pattern of square numbers.		☺	😐	☹
Unit 3 – Reading and writing whole numbers	**Date:**			
• I can read numbers to 100 000.		☺	😐	☹
• I can write numbers to 100 000.		☺	😐	☹
• I can order a set of positive and negative numbers.		☺	😐	☹
Unit 4 – Addition and subtraction	**Date:**			
• I can add combinations of 1-, 2- and 3-digit numbers mentally.		☺	😐	☹
• I can subtract combinations of 1-, 2- and 3-digit numbers mentally.		☺	😐	☹
• I can use mental strategies to add and subtract money.		☺	😐	☹
• I can solve missing number problems, such as $234 + \square = 534$.		☺	😐	☹
Unit 5 – Addition	**Date:**			
• I can use a written method to add a 2-digit number to a 3-digit number.		☺	😐	☹
• I can use a written method to add two 3-digit numbers.		☺	😐	☹
• I can estimate the answer to an addition calculation.		☺	😐	☹

Unit 6 – Subtraction	Date:			
• I can use a written method to subtract a 2-digit number from a 3-digit number.		☺	😐	☹
• I can use a written method to subtract two 3-digit numbers.		☺	😐	☹
• I can estimate the answer to a subtraction calculation.		☺	😐	☹

Unit 7 – Times tables	Date:			
• I know the multiplication facts for the times tables from 1 to 10.		☺	😐	☹
• I know the division facts for the times tables from 1 to 10.		☺	😐	☹

Unit 8 – Multiples, factors and divisibility	Date:			
• I understand and can identify multiples.		☺	😐	☹
• I understand and can identify factors.		☺	😐	☹
• I understand the relationship between multiples and factors.		☺	😐	☹
• I understand tests of divisibility by 2, 5, 10, 25, 50 and 100.		☺	😐	☹

Unit 9 – Multiplication (A)	Date:			
• I can group numbers in different ways to make multiplication simpler.		☺	😐	☹
• I can multiply a tens number by a 1-digit number, such as $60 \times 9 =$.		☺	😐	☹
• I can multiply a hundreds number by a 1-digit number, such as $400 \times 8 =$.		☺	😐	☹
• I can multiply a 2-digit number by a 1-digit number.		☺	😐	☹
• I can estimate the answer to a multiplication calculation.		☺	😐	☹

Unit 10 – Multiplication (B)	Date:			
• I can use a written method to multiply a 2-digit number by a 1-digit number.		☺	😐	☹
• I can use a written method to multiply a 3-digit number by a 1-digit number.		☺	😐	☹
• I can estimate the answer to a multiplication calculation.		☺	😐	☹

Unit 11 – Division (A)	Date:			
• I can divide a 2-digit number by a 1-digit number.		☺	😐	☹
• I can estimate the answer to a division calculation.		☺	😐	☹

Unit 12 – Division (B)	Date:			
• I can divide a 2-digit number by a 1-digit number, including using a written method.		☺	😐	☹
• I can estimate the answer to a division calculation.		☺	😐	☹
Unit 13 – Place value and ordering	Date:			
• I understand the value of each digit in 4- and 5-digit numbers.		☺	😐	☹
• I can compose and decompose 4- and 5-digit numbers.		☺	😐	☹
• I can regroup numbers in different ways.		☺	😐	☹
• I can compare and order numbers, including negative numbers.		☺	😐	☹
Unit 14 – Place value, ordering and rounding	Date:			
• I can use place value to multiply whole numbers by 10 and 100.		☺	😐	☹
• I can use place value to divide whole numbers by 10 and 100.		☺	😐	☹
• I can round numbers to the nearest 10, 100, 1000, 10 000 or 100 000.		☺	😐	☹
Unit 15 – Fractions (A)	Date:			
• I understand that the larger the denominator, the smaller the fraction.		☺	😐	☹
• I understand that fractions can be combined to make one whole.		☺	😐	☹
• I can recognise two proper fractions with the same value.		☺	😐	☹
• I can compare and order fractions.		☺	😐	☹
Unit 16 – Fractions (B)	Date:			
• I understand that a fraction can be shown as a division.		☺	😐	☹
• I understand that fractions can act as operators.		☺	😐	☹
• I can add fractions with the same denominator.		☺	😐	☹
• I can subtract fractions with the same denominator.		☺	😐	☹
Unit 17 – Percentages	Date:			
• I understand that a percentage is the number of parts in 100.		☺	😐	☹
• I can write hundredths as percentages.		☺	😐	☹
• I understand that fractions can be represented as percentages.		☺	😐	☹

Unit 18 – Time	Date:			
• I understand the relationship between units of time, and can convert between them.		☺	😐	☹
• I can read and show times on analogue and 12- and 24-hour digital clocks.		☺	😐	☹
• I can use and interpret timetables in 12- and 24-hour times.		☺	😐	☹
• I can find time intervals between different units of time.		☺	😐	☹
Unit 19 – 2D shapes and symmetry	Date:			
• I can investigate shapes made by combining two or more shapes.		☺	😐	☹
• I can recognise shapes that tessellate and those that do not.		☺	😐	☹
• I can identify horizontal, vertical and diagonal lines of symmetry.		☺	😐	☹
Unit 20 – 3D shapes	Date:			
• I can identify 2D faces of 3D shapes and describe their properties.		☺	😐	☹
• I can identify prisms and describe their properties.		☺	😐	☹
• I can identify pyramids and describe their properties.		☺	😐	☹
• I can match nets to their corresponding 3D shapes.		☺	😐	☹
Unit 21 – Angles	Date:			
• I can identify angles in shapes.		☺	😐	☹
• I can identify and combine right angles.		☺	😐	☹
• I can identify, estimate, compare, order and classify acute, right and obtuse angles.		☺	😐	☹
Unit 22 – Measuring instruments	Date:			
• I can read and interpret scales on a ruler.		☺	😐	☹
• I can read and interpret measuring scales.		☺	😐	☹
• I can read and interpret scales on measuring cylinders.		☺	😐	☹
Unit 23 – Perimeter and area	Date:			
• I can find the perimeter of regular 2D shapes.		☺	😐	☹
• I can find the area of regular 2D shapes and compound shapes.		☺	😐	☹
• I can estimate the area of regular and irregular 2D shapes.		☺	😐	☹

Unit 24 – Position, direction and movement	Date:			
• I can use compass points to describe position, direction and movement.		☺	😐	☹
• I can use coordinates to describe position.		☺	😐	☹
• I can read and plots points on a coordinate grid.		☺	😐	☹
Unit 25 – Position, direction, movement and reflection	Date:			
• I can read and plots points on a coordinate grid for vertices of shapes.		☺	😐	☹
• I can reflect 2D shapes in a horizontal or vertical mirror line.		☺	😐	☹
Unit 26 – Statistics	Date:			
• I can record, organise, represent and interpret data in a tally chart or frequency table to answer a question.		☺	😐	☹
• I can record, organise, represent and interpret data in a Venn or Carroll diagram to answer a question.		☺	😐	☹
• I can record, organise, represent and interpret data in a pictogram to answer a question.		☺	😐	☹
Unit 27 – Statistics and probability	Date:			
• I can record, organise, represent and interpret data in a bar chart to answer a question.		☺	😐	☹
• I can record, organise, represent and interpret data in dot plots to answer a question.		☺	😐	☹
• I can use familiar language associated with chance to describe events and chance experiments.		☺	😐	☹

Unit 1 Counting and sequences (A)

Name: _____

1 Continue each sequence.

a 1771, 1871, 1971, ☐ , ☐ , ☐ , ☐

b 6453, 5453, 4453, ☐ , ☐ , ☐ , ☐

c 8532, 8522, 8512, ☐ , ☐ , ☐ , ☐

d 3895, 4895, 5895, ☐ , ☐ , ☐ , ☐

e 5569, 5469, 5369, ☐ , ☐ , ☐ , ☐

2 Write the missing numbers.

a 3487, 3497, ☐ , ☐ , 3527, ☐ , ☐

b 5436, ☐ , 5236, ☐ , ☐ , ☐ , 4836

c ☐ , 2874, ☐ , ☐ , 5874, ☐ , 7874

d 8422, ☐ , ☐ , ☐ , 8382, ☐ , 8362

e 6701, 6801, ☐ , ☐ , ☐ , 7201, ☐

3 Complete each statement about adding odd (O) and even (E) numbers. Then write a calculation to prove that your statement is correct.

a O + O = ☐

☐ + ☐ = ☐

b E + E = ☐

☐ + ☐ = ☐

c O + E = ☐

☐ + ☐ = ☐

d E + O = ☐

☐ + ☐ = ☐

Number

Complete each statement about subtracting odd (O) and even (E) numbers. Then write a calculation to prove that your statement is correct.

a O – O = ☐

☐ – ☐ = ☐

b E – E = ☐

☐ – ☐ = ☐

c O – E = ☐

☐ – ☐ = ☐

d E – O = ☐

☐ – ☐ = ☐

 a Write the missing numbers.

12, 16, ☐, 24, 28, 32, ☐, ☐, 44

b What is the rule? _____

 a Write the missing numbers.

92, 85, 78, ☐, ☐, 57, 50, 43, ☐

b What is the rule? _____

 a Write the missing numbers.

82, 91, ☐, 109, 118, ☐, 136, 145, ☐

b What is the rule? _____

 a Write the missing numbers.

73, 67, 61, ☐, 49, ☐, 37, 31, ☐

b What is the rule? _____

Now look at and think about each of the *I can* statements.

☐

Date: _____

Number

Name: _____

1 Continue each sequence.

a 3, 12, 21, 30, 39, ☐ , ☐ , ☐ , ☐ , ☐

b 62, 56, 50, 44, 38, ☐ , ☐ , ☐ , ☐ , ☐

c 30, 37, 44, 51, 58, ☐ , ☐ , ☐ , ☐ , ☐

d 73, 65, 57, 49, 41, ☐ , ☐ , ☐ , ☐ , ☐

2 Write the missing numbers.

a 81, 78, 75, ☐ , ☐ , 66, ☐ , 60, ☐

b 48, 43, ☐ , 33, 28, ☐ , 18, ☐ , ☐

c 6, 10, ☐ , ☐ , 22, 26, ☐ , 34, ☐

d 75, ☐ , 59, 51, ☐ , ☐ , 27, 19, ☐

3 Continue each sequence.

a 9, 7, 5, 3, ☐ , ☐ , ☐ , ☐ , ☐

b 5, 4, 3, 2, ☐ , ☐ , ☐ , ☐ , ☐

c 10, 6, 2, –2, ☐ , ☐ , ☐ , ☐ , ☐

4 Write the missing numbers.

a 20, 15, ☐ , ☐ , 0, –5, ☐ , –15, ☐

b 50, 40, ☐ , 20, 10, ☐ , ☐ , ☐ , –30

c 8, ☐ , 2, ☐ , –4, ☐ , –10, –13, ☐

a Write the missing numbers.

−28, −22, ☐, −10, −4, ☐, 8, ☐, 20

b What is the rule? _____

 a Write the missing numbers.

−30, −20, −10 ☐, ☐, 20, 30, 40, ☐

b What is the rule? _____

 a Write the missing numbers.

13, 6, ☐, −8, −15, ☐, −29, −36, ☐

b What is the rule? _____

8 Continue the sequence of square numbers.

1, 4, 9, ☐, ☐, ☐

9 Lee says that 48 is a square number.

a Do you agree? ☐

b Draw a diagram to justify your answer.

Now look at and think about each of the *I can* statements.

☐

Date: _____

Number

Name: _____

1 Draw lines to match each numeral to its number name.

2685 ●   ● two thousand, six hundred and five

12476 ●   ● two thousand, six hundred and eighty-five

2605 ●   ● four thousand, three hundred and ninety-three

4393 ●   ● forty-three thousand, eight hundred and thirteen

43813 ●   ● twelve thousand, four hundred and seventy-six

2 Write the matching numerals.

a two thousand, three hundred and forty-one   [ ]

b forty-one thousand, two hundred and seventy-nine   [ ]

c eight thousand, six hundred and fifteen   [ ]

d sixteen thousand, five hundred and eight   [ ]

e seventy-one thousand, four hundred and fifty   [ ]

3 Write the matching number names.

a 7346   [   ]

b 58183   [   ]

c 5503   [   ]

d 18492   [   ]

e 70687   [   ]

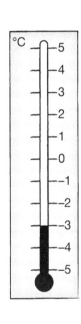

Number

4 Write each of these numbers as a numeral and in words.

a 6 thousands, 4 hundreds, 3 tens, 9 ones.

b 1 hundred, 34 thousands, 7 ones, 0 tens.

5 Write the temperature shown on each thermometer.

a [ ]

b [ ]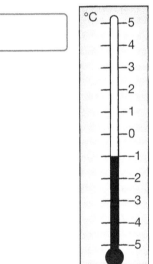

c [ ]

6 Draw an arrow from each number to show where it belongs on the number line.

a 5   −3   −4   3

−10   0   10

b −7   2   −2   6

−10   0   10

Now look at and think about each of the *I can* statements.

[ ]

Date: _____

15

Name: _____

Number

1 Use a **mental strategy** to answer each calculation.
Explain your strategies.

a 213 + 99 = ☐

☐

b 543 + 26 = ☐

☐

c 342 + 136 = ☐

☐

d 265 + 325 = ☐

☐

2 Write the missing numbers.

a 345 + ☐ = 845

b 534 + ☐ = 734

c ☐ + 400 = 624

d 112 + ☐ = 213

3 Use a **mental strategy** to answer each calculation.
Explain your strategies.

a $134 + $98 = ☐

☐

b $399 + $56 = ☐

☐

c $542 + $127 = ☐

☐

d $602 + $288 = ☐

☐

4 Write the missing numbers.

a ☐ + $200 = $367

b $582 + ☐ = $882

c $552 + ☐ = $653

d ☐ + $300 = $499

Number

5 Use a **mental strategy** to answer each calculation.
Explain your strategies.

a 534 – 99 = [ ]

[ ]

b 376 – 45 = [ ]

[ ]

c 756 – 523 = [ ]

[ ]

d 756 – 299 = [ ]

[ ]

6 Write the missing numbers.

a [ ] – 300 = 654

b 619 – [ ] = 219

c 756 – [ ] = 56

d [ ] – 199 = 234

7 Use a **mental strategy** to answer each calculation.
Explain your strategies.

a $445 – $97 = [ ]

[ ]

b $649 – $36 = [ ]

[ ]

c $846 – $515 = [ ]

[ ]

d $753 – $499 = [ ]

[ ]

8 Write the missing numbers.

a $948 – [ ] = $248

b $555 – [ ] = $255

c [ ] – $200 = $121

d $683 – [ ] $482

Now look at and think about each of the *I can* statements.

[ ]

Date: _____

Number

Name: _____

 1 Use a **mental strategy** to answer each calculation.
Explain your strategies.

a 545 + 39 = ☐

b 239 + 41 = ☐

 2 Estimate the answer to each calculation, writing your estimate in the bubble. Then use the **expanded written method** to work out the answer. Show your working out.

a 644 + 67

+

b 426 + 83

+

c 848 + 78

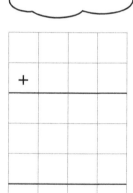

+

 3 Estimate the answer to each calculation, writing your estimate in the bubble. Then use the **formal written method** to work out the answer. Show your working out.

a 347 + 86

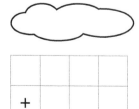

+

b 558 + 77

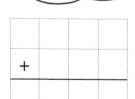

+

c 876 + 45

+

4 Use a **mental strategy** to answer each calculation.
Explain your strategies.

a 435 + 149 = ☐

b 627 + 289 = ☐

 5 Estimate the answer to each calculation, writing your estimate in the
bubble. Then use the **expanded written method** to work out the answer.
Show your working out.

a 546 + 175

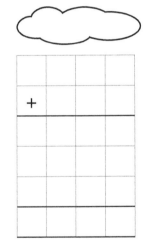

b 357 + 268

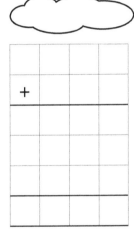

c 768 + 156

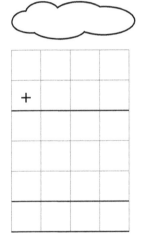

 6 Estimate the answer to each calculation, writing your estimate in the
bubble. Then use the **formal written method** to work out the answer.
Show your working out.

a 446 + 286

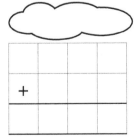

b 693 + 188

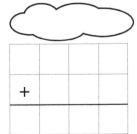

c 465 + 487

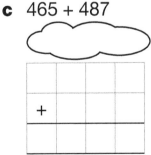

Now look at and think
about each of the
I can statements. ☐

Date: _____

Name: _____

1 Use a **mental strategy** to answer each calculation.
Explain your strategies.

a 675 – 54 = ☐

b 474 – 69 = ☐

2 Estimate the answer to each calculation, writing your estimate in the bubble. Then use the **expanded written method** to work out the answer. Show your working out.

a 534 – 78

b 326 – 57

c 945 – 66

3 Estimate the answer to each calculation, writing your estimate in the bubble. Then use the **formal written method** to work out the answer. Show your working out.

a 413 – 88

b 856 – 77

c 621 – 43

Number

4 Use a **mental strategy** to answer each calculation.
Explain your strategies.

a 843 – 399 = ☐

b 745 – 639 = ☐

5 Estimate the answer to each calculation, writing your estimate in the
bubble. Then use the **expanded written method** to work out the answer.
Show your working out.

a 856 – 269

b 531 – 176

c 422 – 268

6 Estimate the answer to each calculation, writing your estimate in the
bubble. Then use the **formal written method** to work out the answer.
Show your working out.

a 744 – 486

b 365 – 187

c 663 – 496

Now look at and think
about each of the
I can statements.

Date: _____

Name: _____

Number

 Complete each fact.

a 5 × 6 = ☐   **b** 4 × 8 = ☐   **c** 9 × 3 = ☐

d 8 × 8 = ☐   **e** 3 × 5 = ☐   **f** 2 × 10 = ☐

g 4 × 7 = ☐   **h** 10 × 9 = ☐   **i** 1 × 4 = ☐

j 6 × 3 = ☐   **k** 9 × 5 = ☐   **l** 5 × 7 = ☐

m 3 × 2 = ☐   **n** 8 × 1 = ☐   **o** 6 × 4 = ☐

 Complete each fact.

a 12 ÷ 6 = ☐   **b** 10 ÷ 2 = ☐   **c** 54 ÷ 9 = ☐

d 36 ÷ 4 = ☐   **e** 50 ÷ 10 = ☐   **f** 25 ÷ 5 = ☐

g 6 ÷ 1 = ☐   **h** 24 ÷ 8 = ☐   **i** 12 ÷ 3 = ☐

j 48 ÷ 8 = ☐   **k** 16 ÷ 4 = ☐   **l** 100 ÷ 10 = ☐

m 21 ÷ 3 = ☐   **n** 42 ÷ 6 = ☐   **o** 72 ÷ 9 = ☐

 Complete each fact.

a 1 × ☐ = 5   **b** ☐ ÷ 2 = 7   **c** 3 × ☐ = 30

d ☐ ÷ 10 = 7   **e** 6 × ☐ = 36   **f** ☐ ÷ 5 = 8

g 2 × ☐ = 4   **h** ☐ ÷ 8 = 2   **i** 4 × ☐ = 20

j ☐ ÷ 4 = 2   **k** 1 × ☐ = 7   **l** ☐ ÷ 8 = 7

m 5 × ☐ = 45   **n** ☐ ÷ 9 = 7   **o** 9 × ☐ = 81

p 10 × ☐ = 60   **q** ☐ ÷ 3 = 3   **r** ☐ ÷ 6 = 8

4 Explain the relationship between the 2, 4 and 8 times tables.

_____

_____

_____

_____

 5 Explain the relationship between the 3, 6 and 9 times tables.

_____

_____

_____

_____

 6 Complete each fact.

a $7 \times 10 = \boxed{}$ **b** $42 \div 7 = \boxed{}$ **c** $7 \times 3 = \boxed{}$

d $14 \div 7 = \boxed{}$ **e** $7 \times 7 = \boxed{}$ **f** $56 \div 7 = \boxed{}$

g $7 \times 4 = \boxed{}$ **h** $35 \div 7 = \boxed{}$ **i** $7 \times 9 = \boxed{}$

7 Use each set of three numbers to write two multiplication and two division facts.

a 8, 7, 56

> _____

> _____

> _____

> _____

b 6, 9, 54

> _____

> _____

> _____

> _____

Now look at and think about each of the *I can* statements.

Date: _____

Number

Number

Name: _____

1 Write five multiples of 3.

2 Write five multiples of 8.

3 Write five multiples of 7.

4 Circle the multiples of 4.

6	8	10	12	14
22	28	30	34	48

5 Circle the multiples of 6.

3	10	18	20	24
32	40	46	48	54

6 What is a multiple? _____

⑤ _____

_____

7 Write all the factors of 12.

8 Write all the factors of 36.

9 Write all the factors of 28.

10 Circle the numbers that are **not** factors of 20.

1	2	3	4	5
6	10	12	15	20

11 Circle the numbers that are **not** factors of 16.

1	2	3	4	6
7	8	10	12	16

 12 What is a factor? _____

_____

_____

 13 How do you know a number is divisible by 2?

_____

_____

 14 How do you know a number is divisible by 5?

_____

_____

 15 How do you know a number if divisible by 10?

_____

_____

Now look at and think about each of the *I can* statements.

Date: _____

Number

Name: _____

 Draw lines to match the multiplications.

24 × 3 ●   ● 4 × 4 × 3

32 × 3 ●   ● 3 × 9 × 4

18 × 4 ●   ● 6 × 4 × 3

27 × 4 ●   ● 4 × 8 × 3

16 × 3 ●   ● 2 × 9 × 4

 Use **factors** to help work out the answer to each calculation.

a 32 × 5 = [ ]

b 45 × 6 = [ ]

3 Complete each fact.

a 40 × 4 = [ ]   **b** 60 × 3 = [ ]   **c** 80 × 2 = [ ]

d 20 × 6 = [ ]   **e** 60 × 5 = [ ]   **f** 50 × 8 = [ ]

g 70 × 3 = [ ]   **h** 30 × 9 = [ ]   **i** 90 × 7 = [ ]

4 Complete each fact.

a 500 × 6 = [ ]   **b** 300 × 8 = [ ]   **c** 400 × 3 = [ ]

d 900 × 2 = [ ]   **e** 800 × 6 = [ ]   **f** 600 × 4 = [ ]

g 200 × 7 = [ ]   **h** 700 × 5 = [ ]   **i** 400 × 9 = [ ]

26

5 Estimate the answer to each calculation, writing your estimate in the bubble. Then use the **grid method** to work out the answer. Show your working out.

a 73 × 8 =

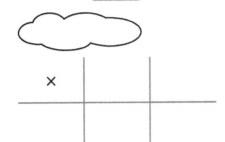

b 58 × 6 =

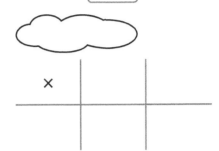

c 44 × 7 =

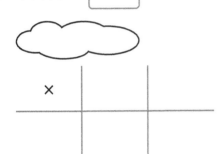

d 69 × 4 =

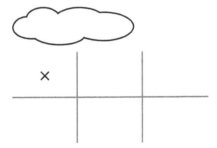

 6 Estimate the answer to each calculation, writing your estimate in the bubble. Then use **partitioning** to work out the answer. Show your working out.

a 36 × 9 =

b 87 × 6 =

Now look at and think about each of the *I can* statements.

Date: _____

Number

Number

Name: _____

1 Estimate the answer to each calculation, writing your estimate in the bubble. Then use the **expanded written method** to work out the answer. Show your working out.

a 74 × 6 = [ ]

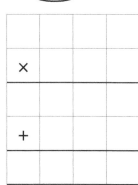

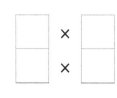

b 58 × 9 = [ ]

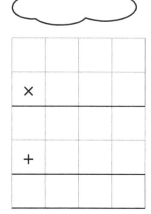

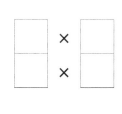

2 Estimate the answer to each calculation, writing your estimate in the bubble. Then use the **grid method** to work out the answer. Show your working out.

a 123 × 6 = [ ]

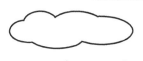

b 136 × 7 = [ ]

c 218 × 3 = [ ]

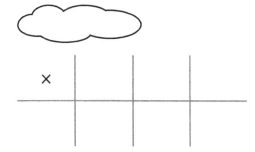

d 232 × 4 = [ ]

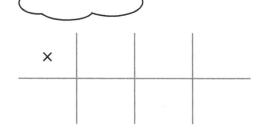

Number

3 Estimate the answer to each calculation, writing your estimate in the bubble. Then use the **expanded written method** to work out the answer. Show your working out.

a 187 × 5 = [ ]

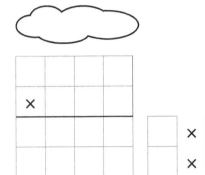

b 276 × 3 = [ ]

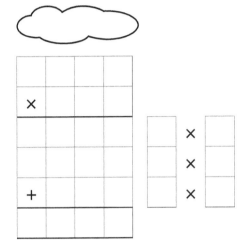

c 242 × 4 = [ ]

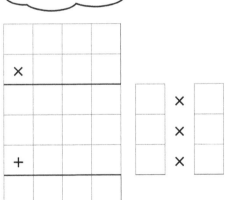

d 118 × 6 = [ ]

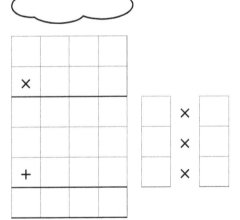

4 Estimate the answer, writing your estimate in the bubble. Then use your **preferred method** to work out the answer. Show your working out.

174 × 8 = [ ]

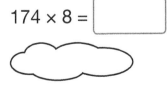

Now look at and think about each of the *I can* statements. [ ]

Date: _____

Name: _____

1 Estimate the answer to each calculation, writing your estimate in the bubble. Then use **partitioning** and known times tables facts to work out the answer. Show your working out.

a 57 ÷ 4 = [   ]

[   ] + [   ] = [   ]

b 64 ÷ 3 = [   ]

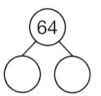

[   ] + [   ] = [   ]

c 92 ÷ 7 = [   ]

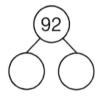

[   ] + [   ] = [   ]

d 97 ÷ 5 = [   ]

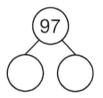

[   ] + [   ] = [   ]

e 87 ÷ 6 = [   ]

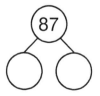

[   ] + [   ] = [   ]

f 98 ÷ 8 = [   ]

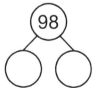

[   ] + [   ] = [   ]

Estimate the answer to each calculation, writing your estimate in the bubble. Then use your **preferred method** to work out the answer. Show your working out.

a 83 ÷ 5 = ☐

b 79 ÷ 4 = ☐

c 88 ÷ 3 = ☐

d 86 ÷ 7 = ☐

e 99 ÷ 8 = ☐

f 95 ÷ 6 = ☐

Now look at and think about each of the *I can* statements. ☐

Date: _____

Number

Name: _____

1 Use the **expanded written method** to work out the answer to each calculation. If you need to, draw a diagram to help you.

a 76 ÷ 4 = [ ]

10	1

b 83 ÷ 3 = [ ]

10	1

c 84 ÷ 5 = [ ]

10	1

d 92 ÷ 6 = [ ]

10	1

2 Estimate the answer to each calculation, writing your estimate in the bubble. Then use your **preferred method** to work out the answer. Show your working out.

a $68 \div 3 =$ ☐

b $89 \div 6 =$ ☐

c $91 \div 4 =$ ☐

d $97 \div 7 =$ ☐

Now look at and think about each of the *I can* statements.

Date: _____

Number

Name: _____

1 Write the number represented on each abacus.

a [   ]   b [   ]

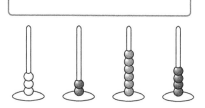

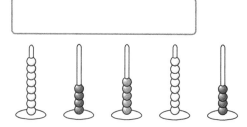

2 Write the numbers represented with place value counters.

a [   ]

1000s	100s	10s	1s
1000 1000 1000 1000 1000 1000 1000	100 100 100 100 100 100	10 10 10 10 10	1 1

b [   ]

10 000s	1000s	100s	10s	1s
10000 10000 10000 10000 10000	1000 1000 1000 1000	100 100 100	10 10 10 10 10 10 10 10	1 1 1 1 1 1 1

3 Write the value of each digit in these numbers.

a 8364 = [   ] + [   ] + [   ] + [   ]

b 72 915 = [   ] + [   ] + [   ] + [   ] + [   ]

c 6834 = [   ] + [   ] + [   ] + [   ]

d 43 821 = [   ] + [   ] + [   ] + [   ] + [   ]

4 In the diagram below, 5748 has been decomposed. Write three other ways to decompose 5748.

[5000 + 700 + 40 + 8]   [   ]

5748

[   ]   [   ]

Number

5 Write four different ways to decompose 6259.

[ ] [ ]

6259

[ ] [ ]

6 Write >, < or = between each pair of numbers.

a 3716 [ ] 4741

b 1327 [ ] 2247

c 6487 [ ] 6478

d 8574 [ ] 8754

7 Write a number that lies between each pair of numbers.

a 4827 [ ] 4836

b 6498 [ ] 6503

c 9581 [ ] 9414

d 7226 [ ] 7248

8 Order each set of numbers, **smallest** first.

a 5734, 4362, 6251, 1298, 3758

[ ] , [ ] , [ ] , [ ] , [ ]

b 7482, 3756, 4872, 5673, 3882

[ ] , [ ] , [ ] , [ ] , [ ]

c 26, −26, 16, −16, −6

[ ] , [ ] , [ ] , [ ] , [ ]

d 6487, 6874, 6784, 6478, 6748

[ ] , [ ] , [ ] , [ ] , [ ]

Now look at and think about each of the *I can* statements.

[ ]

Date: _____

Number

Name: _____

1 Write the matching numeral or number name.

a 867 231 _____

_____

b two hundred and forty-one thousand, eight hundred and seventy-three

[ ]

c one million, eight hundred and fifty thousand, one hundred and twelve

[ ]

2 Write the value of the **bold** digit in each number.

a **7**68 429 [ ]   b 1**2**56 409 [ ]

c **2**685 049 [ ]   d 1**9**58 324 [ ]

3 Multiply each number by 10.

a 576   b 4124   c 23 948

[ ]   [ ]   [ ]

4 Explain what happens when a number is multiplied by 10.

⑤

_____

_____

5 Divide each number by 10.

a 670   b 23 530   c 5580

[ ]   [ ]   [ ]

6 Explain what happens when a number is divided by 10.

⑤

_____

_____

Number

7 Multiply each number by 100.

a 834

b 2897

c 18 746

8 Explain what happens when a number is multiplied by 100.

⑤

_____

_____

9 Divide each number by 100.

a 573 600

b 37 400

c 8500

10 Explain what happens when a number is divided by 100.

⑤

_____

11 Round each number to the nearest 10.

a 63

b 587

c 2745

12 Round each number to the nearest 100.

a 876

b 6834

c 3387

13 Round each number to the nearest 1000.

a 8482

b 17 687

c 27 498

14 Round each number to the nearest 10 000.

a 174 984

b 13 248

c 25 258

15 Round each number to the nearest 100 000.

a 586 448

b 1 258 295

Now look at and think about each of the *I can* statements.

Date: _____

Number

Name: _____

 1 What fraction of each shape is shaded? What fraction is **not** shaded?

a shaded: ▭/▭ not shaded: ▭/▭

b shaded: ▭/▭ not shaded: ▭/▭

 2 Circle the true statements.

$\frac{1}{3} = \frac{2}{6}$ $\frac{2}{4} = \frac{1}{8}$

$\frac{3}{7} = \frac{3}{12}$ $\frac{2}{10} = \frac{1}{5}$

$\frac{3}{4} = \frac{9}{12}$ $\frac{1}{2} = \frac{5}{8}$

 3 Write the missing numerators.

a $\frac{1}{3} = \frac{\Box}{6}$ **b** $\frac{4}{5} = \frac{\Box}{10}$ **c** $\frac{3}{4} = \frac{\Box}{8}$

d $\frac{2}{10} = \frac{\Box}{5}$ **e** $\frac{2}{8} = \frac{\Box}{4}$ **f** $\frac{8}{12} = \frac{\Box}{3}$

4 Circle the **larger** fraction in each pair.

a $\frac{1}{3}$ $\frac{1}{5}$ **b** $\frac{2}{5}$ $\frac{2}{4}$

c $\frac{3}{7}$ $\frac{3}{12}$ **d** $\frac{1}{10}$ $\frac{1}{7}$

e $\frac{5}{6}$ $\frac{5}{12}$ **f** $\frac{1}{9}$ $\frac{3}{9}$

Write the missing fractions.

a $\frac{2}{5} + \dfrac{\boxed{}}{\boxed{}} = 1$

b $\frac{7}{10} + \dfrac{\boxed{}}{\boxed{}} = 1$

c $\frac{3}{7} + \dfrac{\boxed{}}{\boxed{}} = 1$

 Write the missing numerators.

a $\frac{2}{3} + \dfrac{\boxed{}}{6} = 1$

b $\frac{2}{8} + \dfrac{\boxed{}}{4} = 1$

c $\frac{4}{10} + \dfrac{\boxed{}}{5} = 1$

 Write < or > to compare these fractions.

a $\frac{1}{3}\ \boxed{}\ \frac{1}{5}$

b $\frac{2}{5}\ \boxed{}\ \frac{2}{4}$

c $\frac{3}{7}\ \boxed{}\ \frac{3}{12}$

d $\frac{1}{10}\ \boxed{}\ \frac{1}{7}$

e $\frac{2}{3}\ \boxed{}\ \frac{1}{3}$

f $\frac{5}{8}\ \boxed{}\ \frac{5}{6}$

 Order each set of fractions, starting with the **smallest**.

a $\frac{1}{4}$ $\frac{3}{8}$ $\frac{3}{4}$ $\frac{1}{2}$ $\frac{5}{8}$

b $\frac{4}{5}$ $\frac{1}{2}$ $\frac{7}{10}$ $\frac{1}{5}$ $\frac{3}{10}$

Now look at and think about each of the *I can* statements.

$\boxed{}$

Date: _____

Number

Name: _____

1 Write the division calculation and fraction that matches each bar.

a

b

c

2 Write each fraction statement as a division calculation and work out the answer.

a $\frac{1}{6}$ of 18 ⬚ ÷ ⬚ = ⬚

b $\frac{1}{7}$ of 42 ⬚ ÷ ⬚ = ⬚

c $\frac{1}{10}$ of 80 ⬚ ÷ ⬚ = ⬚

3 Work out the answer to each division, then write the matching fraction statement.

a 36 ÷ 4 = ⬚ $\frac{⬚}{⬚}$ of ⬚ = ⬚

b 72 ÷ 9 = ⬚ $\frac{⬚}{⬚}$ of ⬚ = ⬚

c 35 ÷ 5 = ⬚ $\frac{⬚}{⬚}$ of ⬚ = ⬚

Number

 Add the fractions.

a $\frac{3}{8} + \frac{2}{8} = \dfrac{\boxed{}}{\boxed{}}$

b $\frac{5}{12} + \frac{3}{12} = \dfrac{\boxed{}}{\boxed{}}$

c $\frac{4}{9} + \frac{2}{9} = \dfrac{\boxed{}}{\boxed{}}$

d $\frac{7}{10} + \frac{2}{10} = \dfrac{\boxed{}}{\boxed{}}$

e $\frac{3}{7} + \frac{2}{7} = \dfrac{\boxed{}}{\boxed{}}$

f $\frac{1}{6} + \frac{4}{6} = \dfrac{\boxed{}}{\boxed{}}$

 Subtract the fractions.

a $\frac{6}{9} - \frac{2}{9} = \dfrac{\boxed{}}{\boxed{}}$

b $\frac{8}{10} - \frac{5}{10} = \dfrac{\boxed{}}{\boxed{}}$

c $\frac{6}{7} - \frac{2}{7} = \dfrac{\boxed{}}{\boxed{}}$

d $\frac{10}{12} - \frac{3}{12} = \dfrac{\boxed{}}{\boxed{}}$

e $\frac{5}{8} - \frac{2}{8} = \dfrac{\boxed{}}{\boxed{}}$

f $\frac{5}{6} - \frac{1}{6} = \dfrac{\boxed{}}{\boxed{}}$

 Write the missing fractions.

a $\frac{2}{10} + \dfrac{\boxed{}}{\boxed{}} = \frac{7}{10}$

b $\frac{3}{5} + \dfrac{\boxed{}}{\boxed{}} = \frac{4}{5}$

c $\frac{7}{9} - \dfrac{\boxed{}}{\boxed{}} = \frac{4}{9}$

d $\frac{3}{8} + \dfrac{\boxed{}}{\boxed{}} = \frac{7}{8}$

e $\frac{9}{12} - \dfrac{\boxed{}}{\boxed{}} = \frac{5}{12}$

f $\frac{6}{7} - \dfrac{\boxed{}}{\boxed{}} = \frac{4}{7}$

g $\dfrac{\boxed{}}{\boxed{}} + \frac{1}{9} = \frac{5}{9}$

h $\dfrac{\boxed{}}{\boxed{}} + \frac{3}{11} = \frac{8}{11}$

i $\dfrac{\boxed{}}{\boxed{}} - \frac{1}{6} = \frac{4}{6}$

Now look at and think about each of the *I can* statements.

Date: _____

Name: _____

You will need
• coloured pencil

1 Circle the percentages.

$\frac{2}{3}$ 20   2%   16   10% $\frac{1}{10}$

2 Circle the **largest** percentage in each group.

a 12%   22%   20%   2%

b 8%   15%   75%   100%

c 10%   1%   50%   30%

3 Write each percentage as a fraction out of 100.

a 5%   **b** 75% =   **c** 30% =

4 Colour each square to represent the percentage.

a 50%   **b** 25%   **c** 75%

5 The shaded part of each grid represents a percentage. Write the percentage shown.

a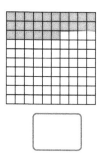

b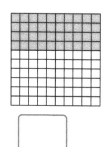

c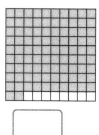

6 Look at each of the grids in **5**. Write the fraction and percentage of each grid that is **not** shaded.

a   =   %   **b**   =   %   **c**   =   %

7 Shade the percentages.

a 12%

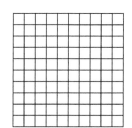

b 56%

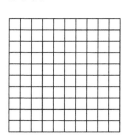

c 79%

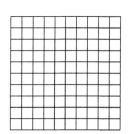

8 Write the equivalent percentage for each fraction.

a $\frac{3}{4}$ = ☐ %

b $\frac{1}{2}$ = ☐ %

c $\frac{1}{4}$ = ☐ %

9 Write the equivalent fraction for each percentage.

a 1% = $\frac{☐}{☐}$

b 33% = $\frac{☐}{☐}$

c 67% = $\frac{☐}{☐}$

10 Write each statement as a fraction and as a percentage.

a 26 out of every 100 $\frac{☐}{☐}$ = ☐ %

b 45 out of every 100 $\frac{☐}{☐}$ = ☐ %

c 78 out of every 100 $\frac{☐}{☐}$ = ☐ %

Now look at and think about each of the *I can* statements.

Date: _____

Name: _____

1 Complete each statement.

a 1 day = [ ] hours

b 1 year = [ ] weeks

c 1 minute = [ ] seconds

d January = [ ] days

e 1 hour = [ ] minutes

f 1 year = [ ] months

g September = [ ] days

h 1 week = [ ] days

2 Draw each time on the analogue clock and write the equivalent digital time.

a | 8 : 43 | a.m. ●
   p.m. ○

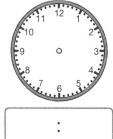

[:]

b | 16 : 51 |

[:] a.m. ○
p.m. ○

c | 2 : 27 | a.m. ○
   p.m. ●

[:]

3 Write the 12-hour and 24-hour digital time to match each analogue clock. The sun or moon shows whether it is a.m. or p.m. 🌙 .

a

[:] a.m. ○
p.m. ○

[:]

b

[:] a.m. ○
p.m. ○

[:]

c

[:] a.m. ○
p.m. ○

[:]

4 Use the timetable to answer the questions.

Kiama to Berry		
Train number	Departs	Arrives
T1	08:23	08:57
T3	10:12	10:35
T5	12:02	12:28
T7	14:53	15:17
T9	16:48	17:15

a What is the journey time of the T3 train?

b How many trains depart Kiama before midday?

c Which train number has a journey time of 26 minutes?

d How long after the T1 train does the T9 train depart?

e Which train number has the longest journey time?

5 Calculate the time intervals.

a 15:06 to 15:47 [   ] minutes

b 03:30 to 10:45 [   ] hours [   ] minutes

c 2nd July, 2020 to 24th July, 2020 [   ] days

d 1st March, 2020 to 1st October, 2020 [   ] months

e 1st January, 2012 to 1st January, 2021 [   ] years

Now look at and think about each of the *I can* statements.

Date: _____

Geometry and Measure

Name: _____

 1 Circle the shapes that are **not** polygons.

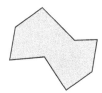 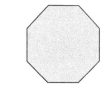  

 2 Name the smaller 2D shapes that have been combined to make each of these larger shapes.

a

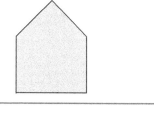

b

c

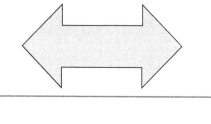

d

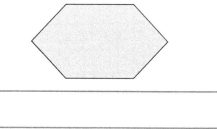

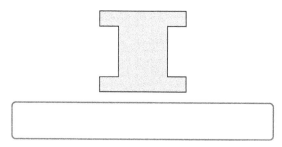 **3** Look at shapes **a** to **d** in **2**. Complete the table.

Shape	Number of sides	Number of vertices	Number of internal right angles
a			
b			
c			
d			

4 Circle the regular shapes that tesselate.

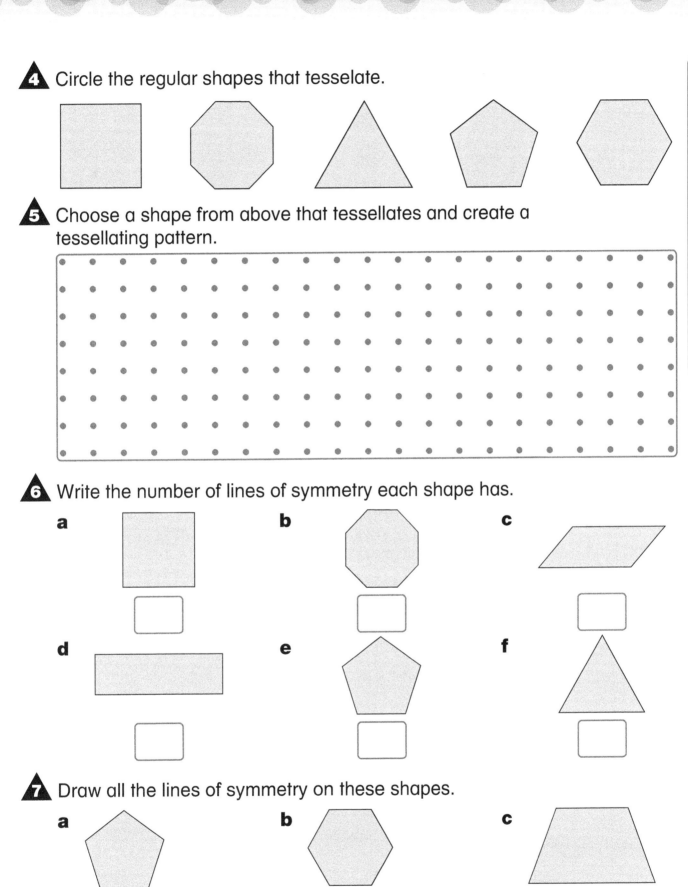

5 Choose a shape from above that tessellates and create a tessellating pattern.

6 Write the number of lines of symmetry each shape has.

a

b

c

d

e

f

7 Draw all the lines of symmetry on these shapes.

a

b

c

Now look at and think about each of the *I can* statements.

Date: _____

Name: _____

Geometry and Measure

1 Sort shapes A to J using your own criteria. Write your criteria at the top of each column in the table. Then write the letter of each shape in the correct column.

A B C D E

F G H I J

2 Look at shapes A to J in **1**. Sort the shapes using a different set of criteria. Write your criteria at the top of each column in the table. Then write the letter of each shape in the correct column.

3 **a** How are shapes C and H in **1** similar? Write two **similarities**.

_____

_____

_____

b How are shapes C and H in **1** different? Write two **differences**.

_____

_____

_____

4 Complete the table.

Shape	Shapes of faces	Number of faces
cuboid		
square-based pyramid		
cylinder		
triangular pyramid		
cube		
triangular prism		

5 Draw lines to match each net to its shape.

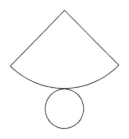

cuboid

cylinder

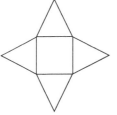

square-based pyramid

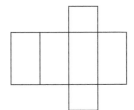

cone

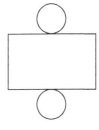

cube

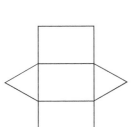

triangular prism

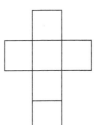

Now look at and think about each of the *I can* statements.

Date: _____

© HarperCollins*Publishers* Limited 2021

Geometry and Measure

Name: _____

1 Mark all the **right** angles in each shape with a cross (**X**).

Mark all the **acute** angles in each shape with an '**A**'.

Mark all the **obtuse** angles in each shape with an '**O**'.

a    b 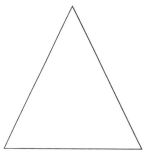   c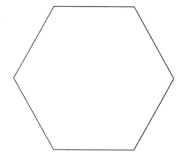

d 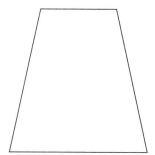   e 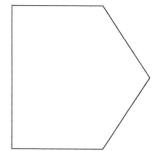   f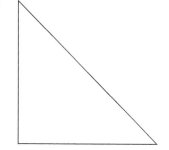

2 Choose an item from the box to complete each statement.

90°	an obtuse angle	180°	half turn
less than a right angle		270°	360°

a A full turn is

b An acute angle is

c An angle greater than a right angle and less than a straight line

is

d A right angle is

e A straight line is

f A straight line is also a

g A three-quarter turn is

Geometry and Measure

 3 **a** Sketch a 4-sided shape with two right angles.

b Describe the other two angles.

_____

_____

_____

 4 Order the angles, from smallest to largest.

a   **b**   **c**   **d**   **e**   **f**

☐   ☐   ☐   ☐   ☐   ☐

 5 Look at the angles in **4**. Write the letters **a** to **f** in the table.

Acute angle	Right angle	Obtuse angle

 6 Draw two different angles in each section of the table.

Acute angle	Right angle	Obtuse angle

Now look at and think about each of the *I can* statements.

☐

Date: _____

Geometry and Measure

Name: _____

1 Complete each statement.

a $\frac{3}{4}l$ = ☐ ml   b 6 m = ☐ cm

c $2\frac{1}{4}$ kg = ☐ g   d 200 cm = ☐ m

e 1500 ml = ☐ l   f 1750 g = ☐ kg

2 Estimate the length of each line.

a _____   ☐

b _____   ☐

3 Use a ruler to measure the length of each line in **2**.

a ☐   b ☐

4 Use a ruler to draw a line 9 centimetres long.

5 Use a ruler to draw a line $10\frac{1}{2}$ centimetres long.

6 Write the mass shown on each scale.

a 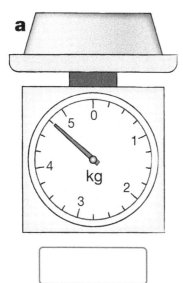   b 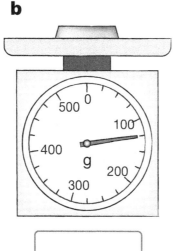   c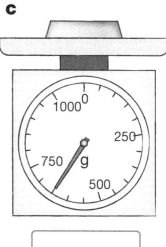

☐   ☐   ☐

Draw an arrow on each scale to show the mass.

a

375 g

b

400 g

c

$2\frac{1}{4}$ kg

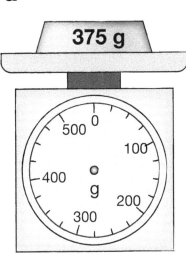

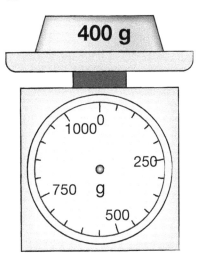

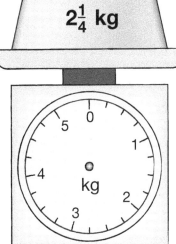

8 How much liquid is in each container?

a   **b**   **c**

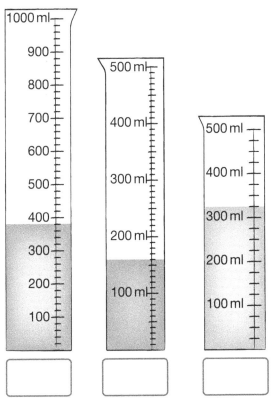

9 Colour the containers to show each amount.

a 260 ml **b** 475 ml **c** 640 ml

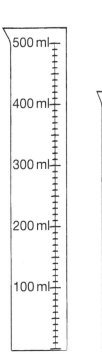

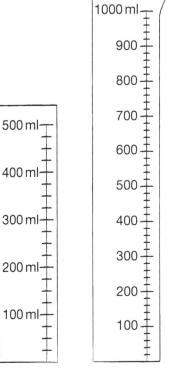

Now look at and think about each of the *I can* statements.

Date: _____

Name: _____

Geometry and Measure

1 Write the perimeter of each of these regular shapes.

a 6 cm

b 7 cm

2 Calculate the perimeter of these regular shapes.
Show your working out.

a Pentagon with sides 6 cm.

b Rectangle with sides 9 cm.

3 What is the area of each compound shape?

= 1 square centimetre (cm^2)

a

b

c

4 Draw a compound shape. Then calculate the perimeter and area of your shape. Show your working out.

= 1 square centimetre (cm^2)

Perimeter:

Area:

5 Calculate the area and perimeter of a rectangle with these dimensions.

a 9 cm × 3 cm

Area: [　　　　　]　　　　　Perimeter: [　　　　　]

b 6 cm × 8 cm

Area: [　　　　　]　　　　　Perimeter: [　　　　　]

c 4 cm × 7 cm

Area: [　　　　　]　　　　　Perimeter: [　　　　　]

 6 Estimate the area of each shape.

[　　] = 1 square centimetre (cm²)

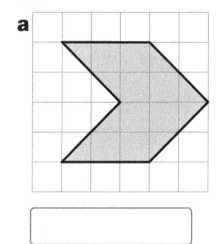

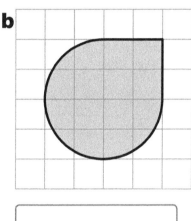

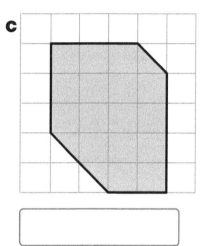

a [　　　　　]　　**b** [　　　　　]　　**c** [　　　　　]

 7 Calculate the perimeter and area of the compound shape. Show your working out.

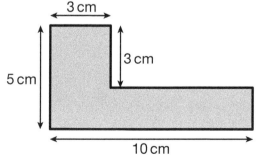

Perimeter: [　　　　　]　　Area: [　　　　　]

Now look at and think about each of the *I can* statements. [　　]

Date: _____

Geometry and Measure

Geometry and Measure

Name: _____

 1 Fill in the missing compass points.

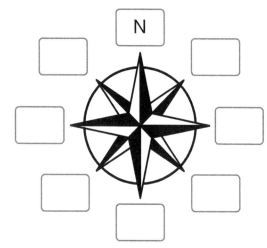

2 Follow the instructions and draw a cross in the square where the frog finishes.

1. 3 squares east
2. 2 squares southeast
3. 1 square south
4. 4 squares west
5. 1 square south
6. 4 squares east
7. 2 squares northeast

3 Write a set of instructions for a route along the white squares from Start to End.

_____

_____

_____

_____

_____

_____

_____

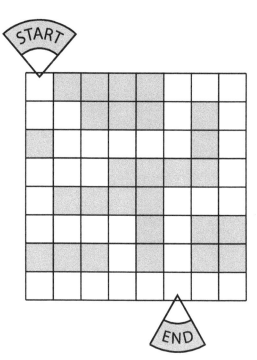

4 Write the coordinates for each letter.

Letter	Coordinates
A	
B	
C	
D	
E	
F	
G	
H	

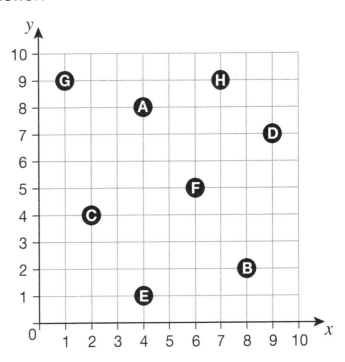

Geometry and Measure

5 Plot points J to Q on the coordinates grid.

Letter	Coordinates
J	(3, 5)
K	(6, 2)
L	(1, 7)
M	(7, 4)
N	(5, 3)
O	(8, 6)
P	(10, 1)
Q	(4, 9)

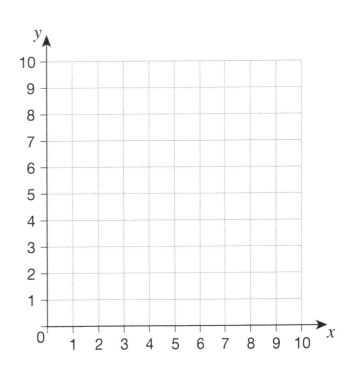

Now look at and think about each of the *I can* statements.

Date: _____

Geometry and Measure

Name: _____

1 Plot the coordinates then join the points to draw a 2D shape. Name each shape.

a (2, 1) (1, 4) (4, 7) (7, 4)
(6, 1)

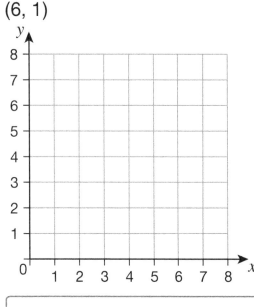

b (3, 7) (5, 7) (7, 4) (5, 1) (3, 1)
(1, 4)

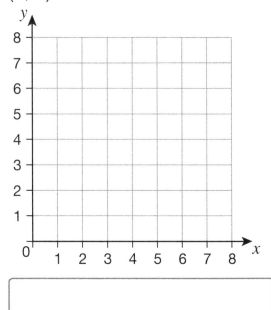

2 Draw an irregular octagon on the grid. Each vertex must be where two grid lines cross. Write the coordinates you used for the vertices of your octagon.

(____, ____)

(____, ____)

(____, ____)

(____, ____)

(____, ____)

(____, ____)

(____, ____)

(____, ____)

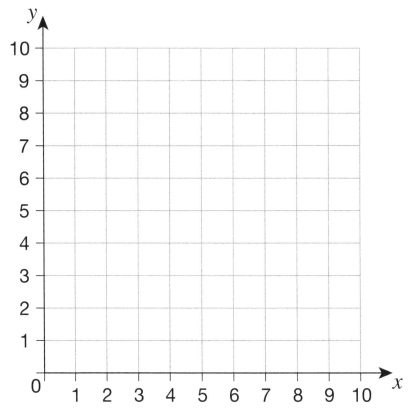

Geometry and Measure

 3 Draw the reflection of each shape in the mirror line.

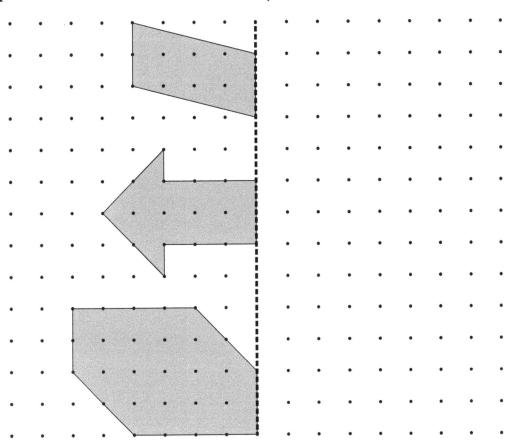

4 Draw the reflection of each shape in the mirror line.

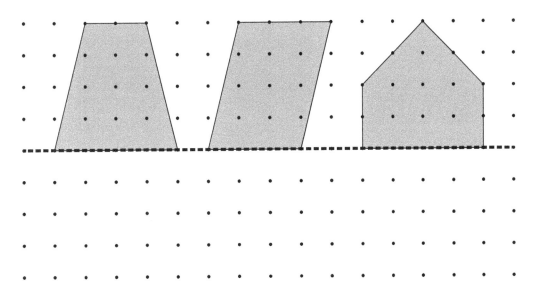

Now look at and think about each of the *I can* statements.

Date: _____

Name: _____

1 Write the numbers in the correct sections of the Venn diagram.

3	4
6	9
12	17
20	25
36	49
52	64

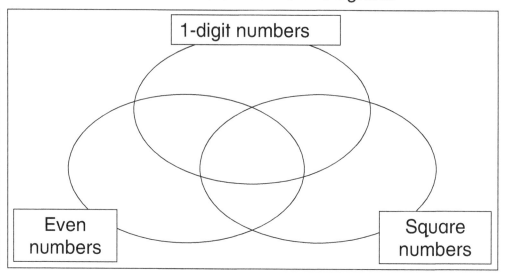

1-digit numbers

Even numbers

Square numbers

2 Write two statements about the data in the Venn diagram in **1**.

_____

_____

_____

3 Look at the 12 numbers in **1**. Decide on two criteria and label the Carroll diagram. Then sort the numbers in the Carroll diagram.

4 Write two statements about the data in the Carroll diagram in **3**.

_____

_____

_____

Statistics and Probability

5 A florist is arranging a display and has made a list of how many flowers she will need to go into different-sized vases. Organise her list in the grouped data frequency table.

1	6	8	10
1	6	8	10
2	6	8	10
3	6	8	11
4	6	8	11
4	6	8	11
4	7	9	11
4	7	9	12
5	7	9	12
5	7	9	14
5	8	9	15

Number of flowers	Tally	Frequency
1–3		
4–6		
7–9		
10–12		
13–15		

6 Use the data in the completed frequency table in 5 to draw a pictogram. Remember to give your pictogram a key. Think carefully about the value that one picture will represent.

Number of flowers

Key:

1–3	
4–6	
7–9	
10–12	
13–15	

7 Write three statements about the data in the frequency table in 5 and in the pictogram in 6.

_____

_____

_____

Now look at and think about each of the *I can* statements.

Date: _____

Name: _____

Statistics and Probability

1 Use the data in the table to draw two bar charts, each with a different scale. Label Bar chart A in intervals of 2 and Bar chart B in intervals of 4.

Stage 4 favourite fruit

Fruit	apple	banana	orange	grapes
Frequency	15	12	10	8

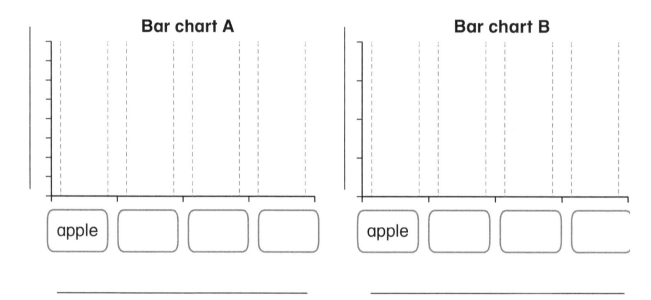

Bar chart A | Bar chart B

apple | | | | apple | | |

_____ | _____

2 How are the two completed bar charts in **1** similar?

_____

_____

3 How are the two completed bar charts in **1** different?

_____

_____

4 Which of the bar charts in **1** do you think is easier to interpret?

Bar chart [ ] Explain your reasons.

_____

_____

_____

5 Use the data in the table to draw a dot plot.

Stage 5 favourite fruit

Fruit	Frequency
apple	10
banana	8
orange	5
grapes	3

apple   banana   orange   grapes

6 Look at the bar charts in **1** and the dot plot in **5**. Write two statements comparing the similarities and differences in the data.

_____

_____

_____

7 Look at this spinner. Choose the correct phrase from the box to describe the chance of spinning:

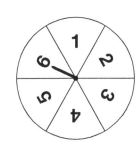

a a number greater than 1 _____

b the number 7 _____

c a multiple of 5 _____

d a number less than 10 _____

e an odd number _____

no chance
poor chance
even chance
good chance
certain

8 Look at your answer to **7 c**. Explain why you think this.

_____

_____

Now look at and think about each of the *I can* statements.

Date: _____

The Thinking and Working Mathematically Star

Think about each of these *I can* statements and record how confident you feel about **Thinking and Working Mathematically**.

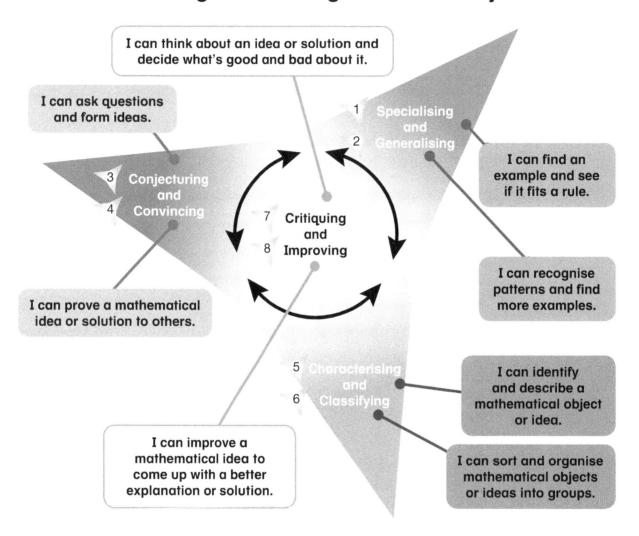

How confident do you feel Thinking and Working Mathematically?

Term ❶	**Term ❷**	**Term ❸**
Date: _____	Date: _____	Date: _____
☺ ☻ ☹	☺ ☻ ☹	☺ ☻ ☹
Date: _____	Date: _____	Date: _____
☺ ☻ ☹	☺ ☻ ☹	☺ ☻ ☹